NORTH GROSVENORDALE BRANCH
THOMPSON LIBRARY
PURCHASED FROM
THE BEQUEST FUNDS
OF
LEOLYN ELIZABETH MOSHER
PHEBE H. CHANDLER

DISCARDED

ANIMAL DEFENSES

ANIMAL DEFENSES

by
Anabel Dean

Illustrated by
Haris Petie

Julian Messner New York

Copyright © 1978 by Anabel Dean
Illustration Copyright © 1978 by Haris Petie

All rights reserved including the right of reproduction in whole or in part in any form. Published by Julian Messner, a Simon & Schuster Division of Gulf & Western Corporation, Simon & Schuster Building, 1230 Avenue of the Americas, New York, N.Y. 10020.

Manufactured in the United States of America

Design by Philip Jaget

Library of Congress Cataloging in Publication Data

Dean, Anabel.
   Animal defenses.
   Includes index.
   SUMMARY: Describes the many ways in which animals avoid their predators.
   1. Animal defenses—Juvenile literature.
[1. Animal defenses]I. Petie, Haris.
II. Title.
QL759.D43   591.5′7   78-17386
ISBN 0-671-32940-5

CONTENTS

1	*Defensive Adaptations*	9
2	*Flight, Weapons, Armor and Others*	11
3	*Protective Coloring*	22
4	*Advertisers*	31
5	*Pretenders*	35
6	*Tricks That Fool*	47
7	*Animal Defenses and People*	55
	Glossary	59
	Index	62

MESSNER BOOKS
BY ANABEL DEAN
AND HARIS PETIE

Animal Defenses
How Animals Communicate
Animals That Fly

ANIMAL DEFENSES

Chapter 1

Defensive Adaptations

In nature all living things eat and are eaten by others. Animals get nourishment from eating plants or other animals. This is the *food chain*, which starts with plants and usually ends with the *carnivores* or meat eaters.

A typical food chain could start with a polliwog eating the green plants growing in a pond. A fish comes along and gobbles up the polliwog. A heron wading in the shallow water swallows the fish. Later a bobcat kills the heron and eats her.

At almost every link of the food chain, animals are catching other animals so they can eat and stay alive. The animal doing the catching is the *predator*. The one who is caught is the *prey*. Many animals are both predator and prey.

But animals don't want to be the prey. All animals want to stay alive. Each animal has some way to defend itself against predators. Defense may be anything from hiding to having natural built-in weapons other animals fear.

It is easy to see how animals with sharp teeth and claws defend themselves. But there are other defenses. Over the centuries, each *species*, or kind of

animal, has slowly changed to give it more protection. This is known as *defensive adaptation.* Some species could not *adapt*, or change, and have died out.

Adaptation comes about through many slow changes in a species. Now and then some animal is born with a new characteristic which makes it easier for it to survive. Perhaps it has longer legs or is slightly different in color. If this new characteristic gives an animal a better chance to survive, it will probably live long enough to have offspring and pass this characteristic on to some of them. Those without this new characteristic are more apt to be caught and eaten before they reproduce. So over the centuries, this species gradually acquires a defensive adaptation to help it survive.

By comparing *fossils,* the hardened remains of the ancestors of some of our present-day animals, we can see how defensive adaptations have changed them. About 55 million years ago, horses were no larger than foxes. But as years passed, those with longer legs, who could run a little faster, had a better chance to survive and reproduce. Horses gradually became much larger, with longer legs.

Many of these defensive adaptations are strange and wonderful. Let's see how they help animals to defend themselves.

Chapter 2

Flight, Weapons, Armor And Others

FLIGHT AS DEFENSE

Most animals try to escape from danger.

Deer, antelope, gazelles and zebras depend largely on the speed of their long legs to escape danger. Often they run in a zigzag fashion so they will be harder to catch. A few animals, such as kangaroos and rabbits, hop instead of run.

And a few birds, such as the ostrich and the roadrunner, also run to escape danger. But most birds fly when threatened. So do bats and many insects.

Wild sheep and goats depend on their sure footedness to get away from pursuers. They climb to places other animals can't reach.

Climbing a tree is a good defense, if the hunter can't climb too! This method of escape is used by bear cubs, squirrels, and many cats and *primates*, such as monkeys and apes.

Swimming is the first defense for some animals who live near water. Frogs jump for the pond when they sense danger and so do some reptiles and birds. Fish and aquatic mammals, of course, use this defense. Beavers, otters and polar bears are some of the strong swimmers among the mammals.

HERDING OR GROUPING AS DEFENSE

Many animals live in herds or packs. The members warn each other of danger and protect each other. Musk oxen form a circle if attacked. All face out, with the young animals inside the circle.

Bees, hornets and ants live in groups. If their nest is attacked, they join together to fight off an attacker. Ants have special soldiers to do their fighting.

HIDING AS DEFENSE

For many animals the best defense is a place to hide. Some animals, such as beavers and many birds, build one. The ovenbird builds a nest inside a rock-hard ball of sand and cow dung. Female hornbills wall themselves and their eggs up inside a hollow tree, leaving only a small hole open. The male feeds the female and young while they are hiding from predators.

Many animals dig a hole and live underground for safety. Others make or use a hole in a tree. The young of some species of fish take shelter inside their parent's mouth.

WEAPONS FOR DEFENSE

TEETH

When we think of animal weapons, we think first of teeth. Many carnivores have long daggerlike canine teeth in both upper and lower jaws. The canine teeth are necessary to catch prey, but are also used to avoid becoming prey. The tusks of the male walrus and the wild pig are long canine teeth. The two upper front teeth—*incisors*—of elephants developed into tusks. Tusks are defensive weapons which few animals care to challenge.

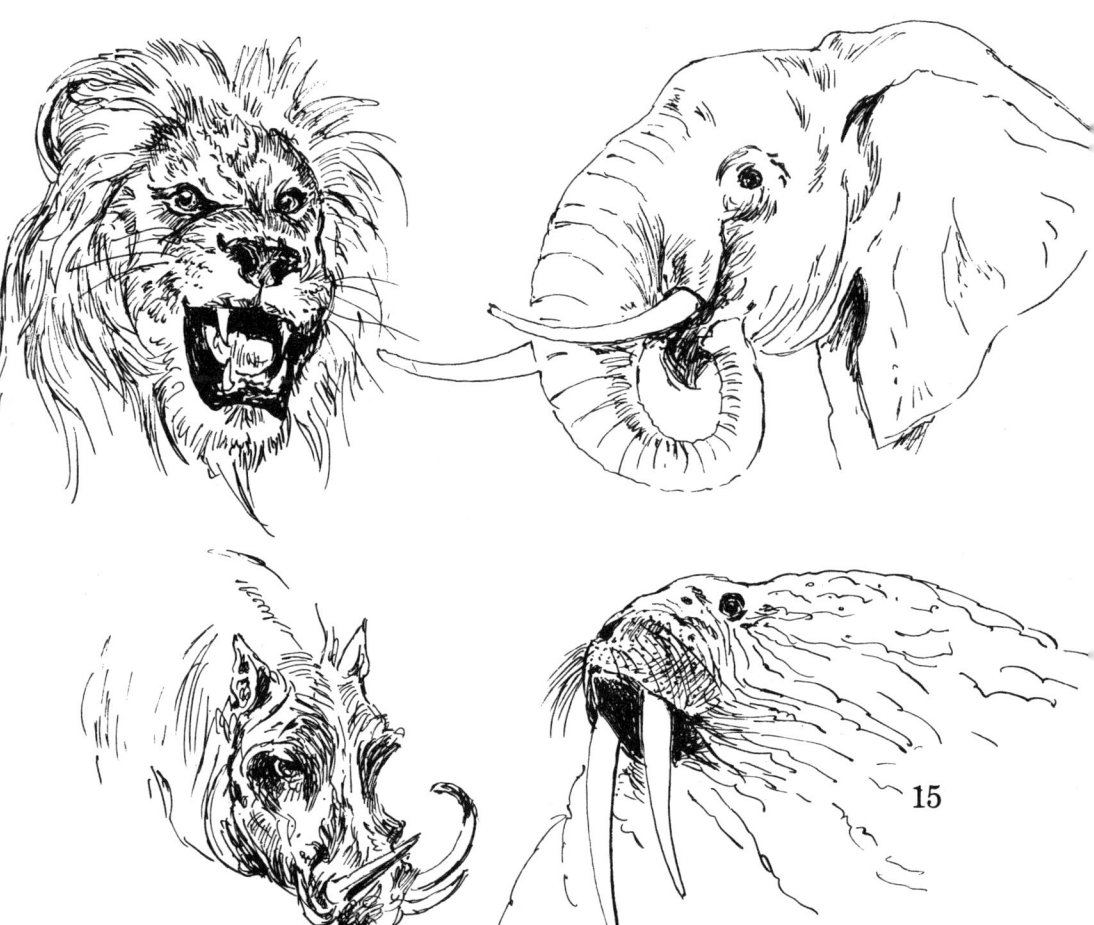

HORNS

The horns of animals like antelope, goats, buffaloes, and musk oxen are used for defense. Both sexes of these animals have horns, which continue growing throughout their lives. Antlers, which are a kind of horn, are shed and regrown every year. Only the males of most deer species have antlers. But female caribou and reindeer as well as male have antlers. Antlers are more for display than are horns, but they are used for defense when necessary.

Carnivores have large claws to help bring down their prey. But those claws are also used as defensive weapons.

Many of the fast runners have hard hooves. If cornered, they use them for defense. Some, such as the horse, kick with their hind feet. Others, such as the deer, rear up and strike with their sharp front hooves.

VENOM

Venom is produced by special glands within an animal's body. Venomous animals have some way, such as stinging or biting, of introducing their venom into another animal.

Among the insects, bees, hornets and ants are venomous. In the ocean, stingray, scorpion fish and jellyfish are venomous. Scorpions and a few spiders have a deadly venom. But the greatest number of venomous animals are sea and land snakes. Venom is a defensive weapon which is feared by both humans and animals.

There are also animals whose bodies are poisonous if eaten. They lack venom-producing glands or a means of injecting poison. The skins of most toads and salamanders secrete a poisonous fluid that discourages predators.

SPECIAL WEAPONS

There are some special defensive weapons used by only a few animals. The electric eel and electric catfish give off a stunning shock of electricity. Cuttlefish, octopuses and squid discharge an inky chemical when threatened. This acts as a "smoke screen," but also dulls the senses of the predator so that it can't pursue.

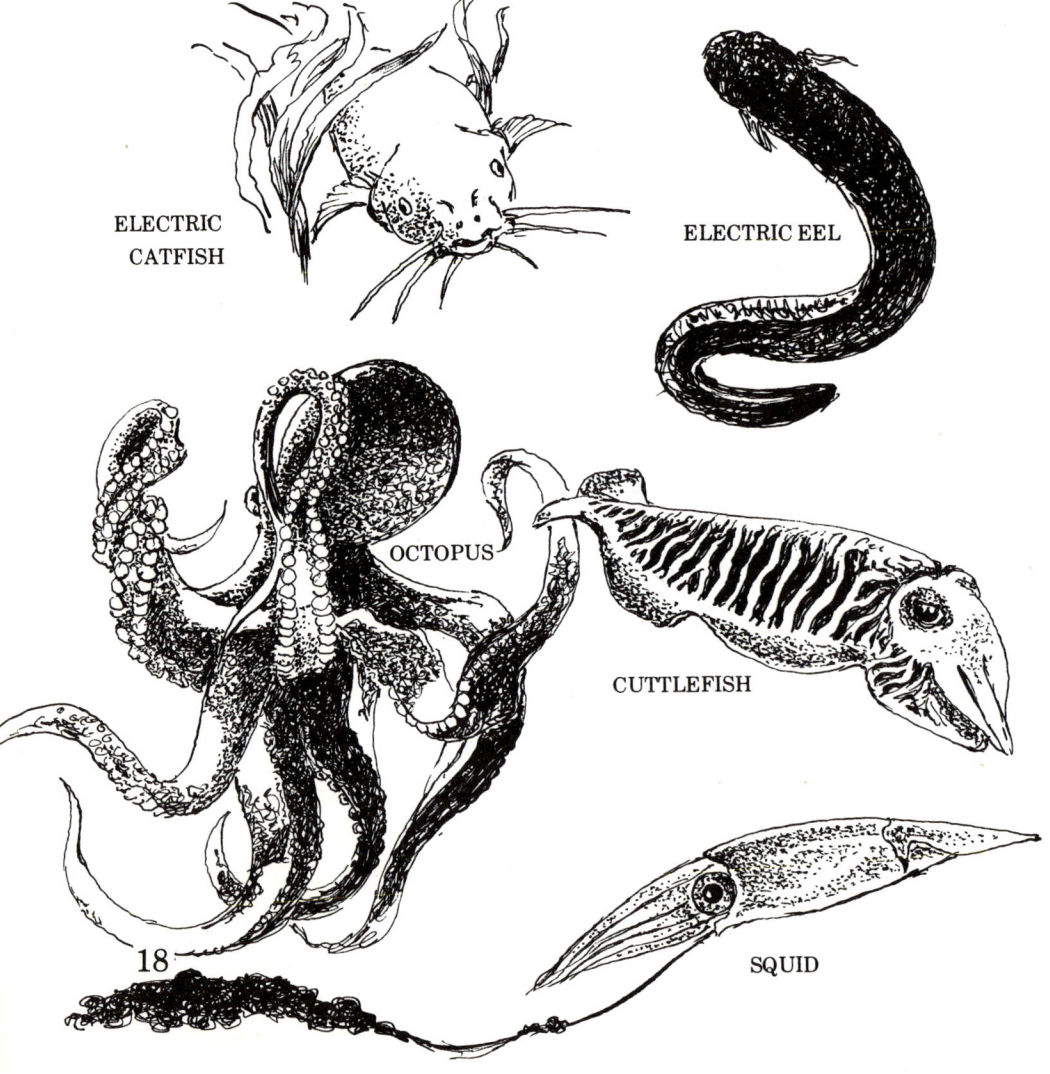

ARMOR FOR DEFENSE

A great variety of animals are protected by armor. The hard covering of many insects' bodies gives them some protection. Some, such as grasshoppers and beetles, also have hard wing coverings to protect their flying wings when at rest.

Crabs and lobsters are among the sea animals with shells covering their entire bodies. Clams, snails, barnacles, oysters and many others pull their soft bodies inside their hard shells for protection. Sea robins and catfish are two fish protected by tough scales and fins. As the hermit crab doesn't have a hard shell of its own, it uses a discarded seashell. When it outgrows one, it finds a larger one.

Some lizards, crocodiles and alligators are armor-plated. The South African armadillo lizard is covered with spiked scales. When threatened, it holds its tail in its mouth to form a circle to protect its soft belly. Turtles withdraw head and feet into their shells when they sense danger.

Armadillos and pangolins are the only mammals with true armor plating. When attacked, some of these animals roll themselves up into a ball. Others press themselves to the ground to protect their bellies, which usually do not have armor plating.

The thick hide of elephants and rhinoceroses acts something like armor to protect them.

Porcupines don't have armor but their quills serve the same purpose. When threatened, the porcupine raises its quills and turns to face an attacker. Often this is enough of a defense. These quills are feared by other animals as the barbed tips can work themselves into the victim's flesh and cause death.

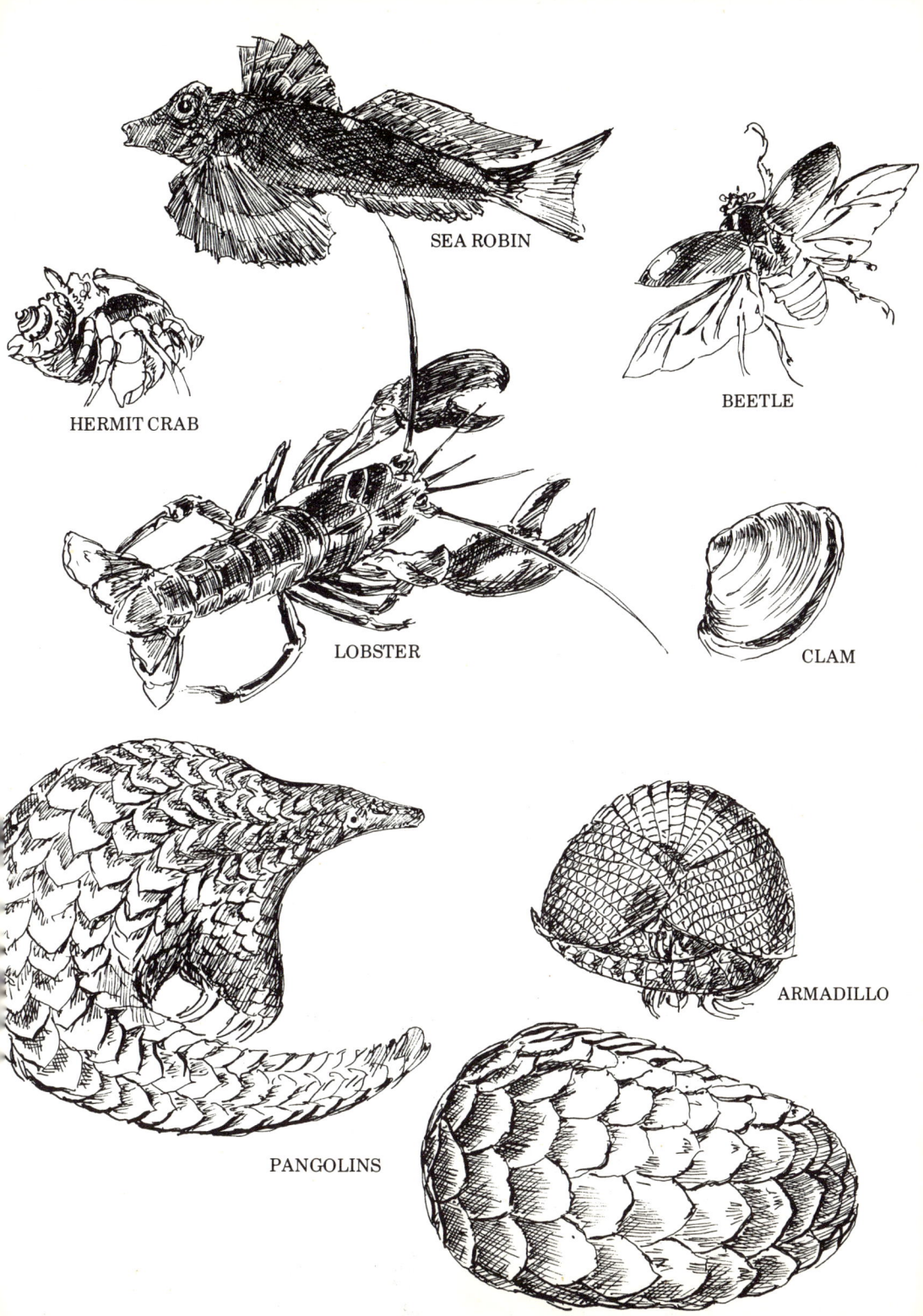

Chapter 3

Protective Coloring

BACKGROUND MATCHING

When the color and markings of an animal's coat match its usual background, it is harder to see. Young animals whose coats match their background are taught to "freeze" when they sense danger. The young sometimes lack the typical scent of their species, so the predator can't smell them either—another defense for the young.

Deer and many other animals have drab brown coats which blend in with the ground, dried grasses and shadows around their habitat. Some insects, snakes, frogs and lizards are green to match the foliage about them. The sloth has a brownish coat to match the trunk and branches of the trees it feeds in. It is almost invisible as it crawls along branches in an upside down position. Algae growing on its fur, as well as on the trees, help it blend into the background.

Three species of anole lizards often hunt together. Each species is a different color to match its usual background. They are easily seen as they scurry about over different vegetation. But when frightened, each species hurries to place itself on a background which matches its color.

Background matching also helps camouflage birds' eggs. Most eggs are colored, or speckled with color, to blend into the nesting site. The sand grouse of the Sudan lays pink-colored eggs under the tree known as the camel's foot tree. When the leaves of this tree fall, they turn the same shade of pink as the grouse eggs, making them invisible.

The eggs of a bird of India are even stranger. The female scoops out a small depression in the dirt, and lays her eggs there. Whatever the color of the soil, even if it is flecked with different colors, the eggs always match the soil.

An Italian naturalist, A. P. di Cesnola, made an experiment to see if protective coloring helped animals survive. He collected 100 praying mantises. Some of them were green and some were brown. He tied threads on them and tethered them to plants so they couldn't crawl away. Twenty green mantises were tied to green plants, and twenty brown mantises to brown plants. The rest of the green mantises were tied to brown plants, while the brown mantises were tied to green plants. Within eighteen days, the sixty mantises whose colors did *not* match their background had been picked off by birds. All of the forty mantises whose colors matched their background had survived.

DISRUPTIVE COLORING

Often the best protective coloring is a spotted, striped or mottled coat. If you see one of these animals away from its natural habitat, it stands out. But in its own environment it blends into the background.

Ocelots, jaguars, servals, leopards and other jungle animals spend most of their time among trees. Their spotted coats blend into the sun-and-shade patterns there.

A tiger's stripes blend into the tall grass. The giraffe's blotchy coat makes it difficult to see under a tree as it feeds on its leaves. The black and white striped coat of the zebra stands out when you see it in the zoo. But at dusk, when predators usually hunt for it, the zebra blends into the background of its home ground.

The young of some animals have different coloring from their parents. Fawns, puma kittens and others are spotted to help them hide, as they wait for their mothers to come back from the hunt for food.

The mottled markings of alligators, lizards and turtles blend into their native habitat. A floating alligator looks like a half-sunken log.

Many fish are striped to blend in with the water plants. In the tropics, fish are often brilliantly colored with splotches of orange and bright blue, making them less visible in the brightness of the sun-lit waters.

Some animals have a few large markings of another color, often around the face or eyes. These splotches break up the outline of their faces and help keep them from being recognized as animals.

The vervet monkey has a light-colored coat and a dark face. Its dark eyes blend into the dark face. A predator often doesn't recognize this dark spot as being an animal's face. The Malayan tapir has large splotches of color which blend into the sun and shadows of the forest where it lives.

COUNTERSHADING

Hunters look for the rounded outline of an animal as it crouches trying to hide. If an animal looks flat, a predator has trouble recognizing it.

You recognize a ball because you see the rounded shape. The light striking the top of the ball makes that part lighter. The lower part, shadowed, looks darker. If this were reversed, with the ball dark on top and light on bottom, it would look flat. This is countershading.

You can see how countershading works. Paint the top of a ball a darker shade than the sides and put it under a good light. Does it look like a ball now?

The fur of most animals is darker on the back and lighter on their sides and bellies. This countershading makes them look flat as they crouch on the ground.

Most fish are also countershaded—darker on top and lighter on their bellies. From above, their dark backs blend into the dark water below. From below, their silvery bellies blend into the lighter water and bright sky of the surface.

But there is one strange catfish of the Nile River whose back is silvery and whose belly is dark. It is still countershaded, however, because this topsy-turvy catfish always swims belly side up. It is the only fish in the world known to swim this way.

A few other animals who normally live in an upside-down position also have reverse countershading. Some spiders, who hang upside down on their webs, have lighter backs and darker bellies.

Animals seem to instinctively realize that countershading helps them hide. One scientist made an experiment with a group of caterpillars. These caterpillars had the usual countershading of darker backs and lighter bellies. They always crawled along the top of twigs and leaves as they fed. The scientist placed a strong light on the ground below the feeding caterpillars. Now their countershading did not help them hide. They all showed up clearly. The caterpillars all promptly crawled to the underside of stems and leaves so their countershading matched the position of the light.

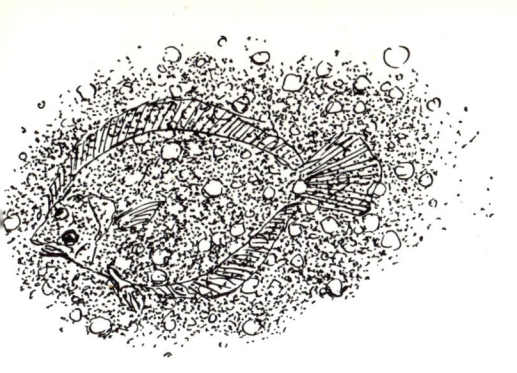

COLOR CHANGES

Several fish change colors to match the ocean bottom where they usually stay. Flounders change colors to match sand and mud, and even become mottled to match a gravel bottom. Squid, cuttlefish and octopuses also change color very rapidly. Chameleons (kuh-mee-lee-unz) and some geckos change color to match their backgrounds.

Some animals change color slowly. A crab spider, who lives on flowers, changes color with the season. In the spring it lives on white flowers and is white. Later in the year when goldenrod are in bloom, it turns gold. A grasshopper of the Russian steppes is bright green in early summer but gradually turns a brownish color as the grass turns brown. The pups of the harp seal are white to blend into the ice where they are born, but turn dark before they take to the water.

Animal colors can also change over many years. A peppered moth, white with black specks, lived in Manchester, England, before 1850. Manchester is an industrial city with many factories pouring out tons of smoke and soot. Even the trunks of the trees here became blackened. By 1950, 90% of the peppered moths in Manchester were black with white specks.

PEOPLE USE CAMOUFLAGE

People have learned about camouflage from studying the way animals hide. During wars, camouflage is used to break up outlines and so make tanks, guns, ships and men harder to see. Men fighting in the jungles wore clothes with splotches of color on them. They covered their helmets with leaves and branches to help them blend into the jungle. Men on Arctic patrol wore all white.

Chapter 4

Advertisers

WARNING COLORATION

Some animals are not eaten by predators. Perhaps they are poisonous or have painful or deadly stings or bites. Maybe they are so distasteful or smelly that no animal wants to eat them. These animals advertise their species with bright colors or distinctive markings so other animals will leave them alone.

The colors most often used as warnings are bright red, yellow, and orange, combined with black, for animals active during the daytime. Many species of bees and wasps use yellow and black stripes. Animals avoid all insects with markings this color.

Warning colors don't save all of these animals. Sometimes a young or inexperienced animal tries to eat one. But after a painful sting or a mouthful of something nasty tasting, it never again tries to eat an animal that looks like that.

Caterpillars of the beautiful orange and black monarch butterfly feed on milkweed plants. This gives both the larva and the butterfly a bitter taste, and predators avoid them. The red and black harlequin cabbage beetle is another distasteful bug which is not eaten.

Many frogs, toads and salamanders have poison-producing glands on their skins. They advertise this fact with loud colors so predators know to stay away. The gold frog of Panama does it with its bright yellow and black coat. A Nicaraguan frog advertises with its eye-catching blue and red coat.

People as well as animals recognize and avoid the deadly coral snake with its red and yellow rings. The black and pink blotches of the poisonous Gila monster also make it easy to see and avoid.

HOVERFLY

BEE

BEETLE   ANT

MIMICRY

Some animals survive by imitating a dangerous or inedible animal. In order to fool predators, they must imitate the animal's actions.

Hover and flower flies are both mimics of bees and wasps. Those that imitate bees not only look like them, but produce a buzz when flying. Those that imitate wasps, mimic their flight pattern. If touched, some of these mimics will bend their abdomens around and jab with it as if they were going to sting.

Most insect-eaters avoid ants because many ants sting or release formic acid, a venom. Flies, beetles, moths, grasshoppers, wasps and spiders all mimic ants. Ants have a very thin midsection, and many mimics have a color pattern that suggests a thin waist. One Sudan grasshopper has the markings of a narrow waist and large abdomen in black on its lighter body.

The leaf-cutting ants of the tropics cut leaves and carry them underground for use in their fungi gardens. In Guyana, there is a bug whose body looks like the leaf carried by these ants. Its legs look like those of a leaf-cutting ant sticking out from under the leaf it is carrying.

MONARCH BUTTERFLY

VICEROY BUTTERFLY

The orange and black monarch butterfly has several look-alikes. The most successful of these mimics is the viceroy butterfly. It is often difficult for even scientists to tell which one has been caught.

The hummingbird moth avoids predators by imitating the hummingbird's actions and looks. Its fuzzy legs and body look like feathers, and its long tongue like the hummingbird's long bill.

HUMMINGBIRD

HUMMINGBIRD MOTH

Chapter 5

Pretenders

Human beings have learned that things are not always what they appear to be. But animals believe what their eyes see. Over the centuries some animals have acquired characteristics which make them look like something *inedible*, not suitable for food. An animal that looks like a leaf, twig, stem, flower or vine has a better chance of surviving.

PRETEND LEAVES

Masquerading as a leaf is one of the commonest disguises of butterflies and moths. When a butterfly alights on a plant, it folds its wings together in an upright position. Now the bright colors of the upper wings are hidden, and only the drab underwings can be seen. The veins of the wings look like the ribs in a leaf and the notched edges like the edges of the leaf. Some insects have markings which resemble a leaf's midrib.

The Kallima butterfly of India, sometimes called the dead-leaf butterfly, is yellow and deep blue. In flight it catches the eye with its brilliant colors, but when it alights and folds its wings, it seems to disappear. The back part of each wing tapers down to look like a narrow stem. This narrow part, or "stem," is rested against a branch so the wings of the butterfly appear to be a dead leaf growing from it. White spots on the underside of the wings look like mildew or fungi patches on a dead leaf.

The walking-leaf beetle wears one of nature's best leaf disguises. Its green body is veined and shaped like a leaf. The wings are also shaped like the two halves of a leaf. Even its legs look like leaflets which have been chewed by caterpillars. When it is windy, the walking-leaf beetle hangs by two legs from a stem and rotates its body so it appears to be windblown. The eggs of the walking-leaf are disguised as seeds, and the larvae like the tiny red buds of the plants they live on.

Several freshwater fish pretend to be leaves. One of the best imitators is the leaf fish, *peche de folha,* (pay-chay day fol-ha) which lives in the tree-lined streams that lead in to the Amazon River. This mottled dark brown fish, with markings to resemble a leaf, matches the dead leaves floating on the water. Its body is extremely thin, with both head and tail coming to a point. It drifts motionlessly, its head and tail drooping down like a limp leaf. A beardlike piece of flesh growing from its chin looks like the leaf's stem. Sometimes the *peche de folha* sinks to the stream bottom to lie among decayed leaves. This disguise not only helps it hide, but enables it to catch smaller fish who have no reason to fear a "dead leaf." If caught in a net, the fish remains motionless and is often thrown back into the water, mistaken for a dead leaf.

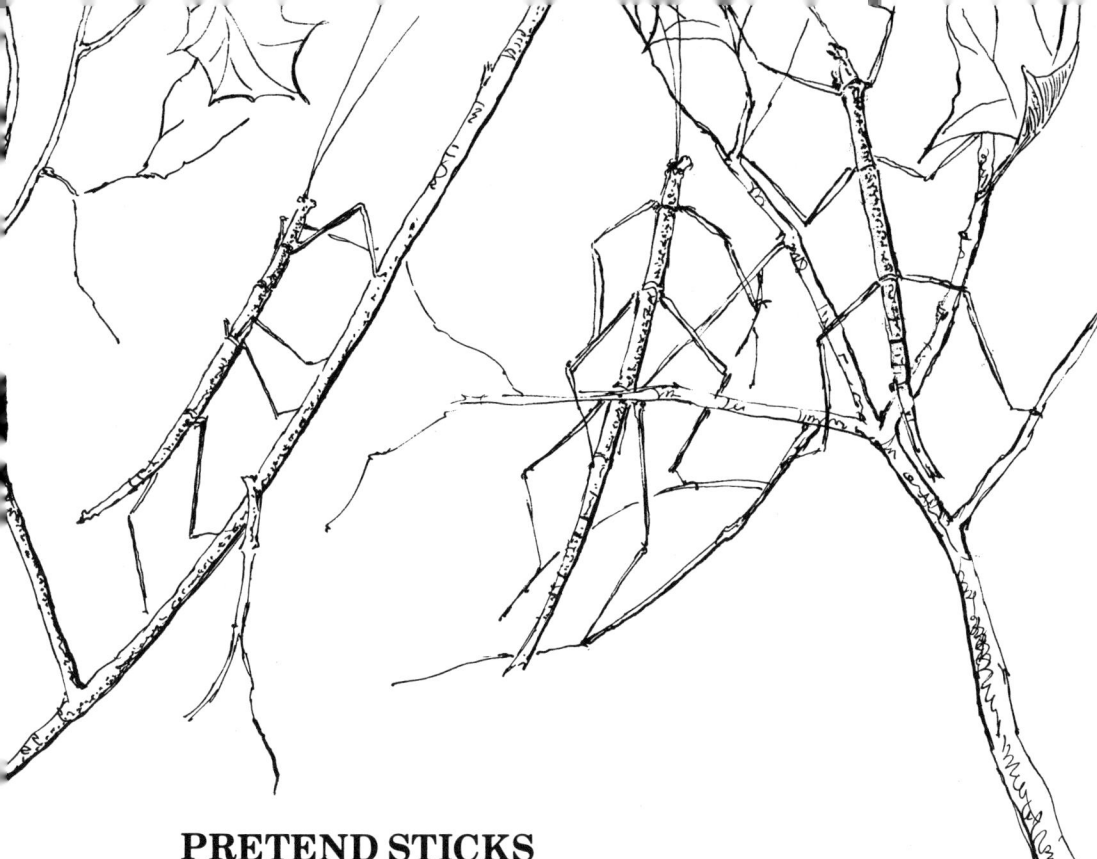

PRETEND STICKS

Because sticks, stems and twigs aren't very appetizing, many animals imitate them. The praying mantises with their sticklike bodies and legs are one of the best known. They move very slowly, which helps keep up the illusion. This disguise helps them get close enough to capture insects.

The walking stick, another insect, also has a long slender sticklike body and legs. When frightened, it "freezes" and remains motionless for a long time. In the spring walking sticks are greenish, but gradually their color changes to brown to match the changing colors of the plants. Walking sticks never leave the plants where they feel safe. Eggs are just dropped to the ground to hatch among the fallen leaves.

Caterpillars also imitate stems or twigs. When the green inchworm senses danger, it clings to a stem with its rear legs and holds its body out stiffly like a small twig. It holds this position for hours if necessary. Night-feeding inchworms pretend they are twigs while they sleep all day. As this position gets tiresome, they tie their upper body to the stem with an invisible thread for support.

PRETEND VINES

Several *Oxybelis* or vine snakes of Central and South America escape their enemies and stalk chameleons by pretending to be vines. Vine snakes can be green, gray or brown depending on the colors of the vines in their habitat. When frightened, the vine snake "freezes" with its body held out horizontally from the plant. If there is a breeze, the snake sways back and forth, just like the vine. Vine snakes are unbelievably slender. When they are full-grown, they are usually about four feet long but only a quarter inch in diameter.

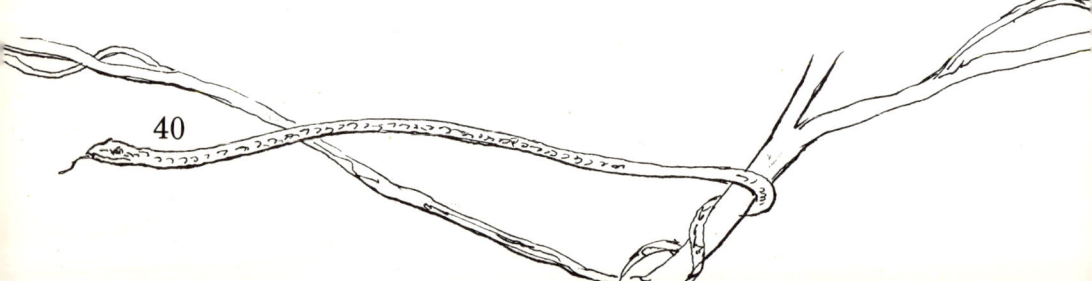

PRETEND SEAWEED

One of the most amazing pretenders is the sargassum fish which lives in the Sargasso Sea. The Sargasso Sea is an area of the Atlantic Ocean between Bermuda and the Azores. Sargassum is a floating brownish seaweed. Sargassum fish do little swimming. They climb through the thick seaweed, using their fins as legs. Even when you hold one in your hand, it is hard to believe it is not a piece of seaweed. Its body even has many leaf and bladderlike growths which match the seaweed.

Another fish, the sea dragon, makes its home in the red seaweed along the Great Barrier Reef off Australia. It is a larger member of the sea horse family. A grown sea dragon is ten or more inches long and the same color as the bright red seaweed it lives in. Leaflike growths from its head, back and tail match the fronds of the seaweed. The sea dragon wraps the end of its tail around a stalk of seaweed and appears to be growing there.

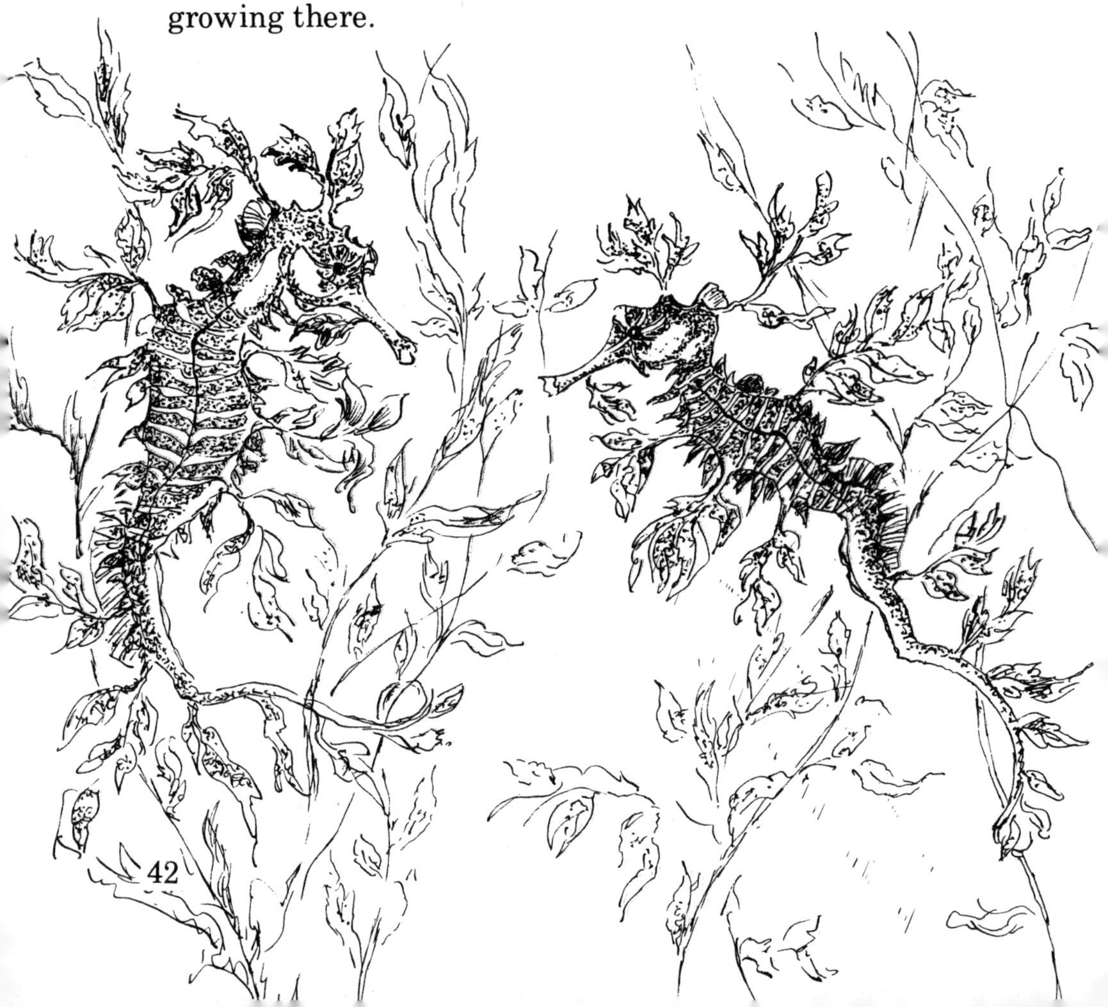

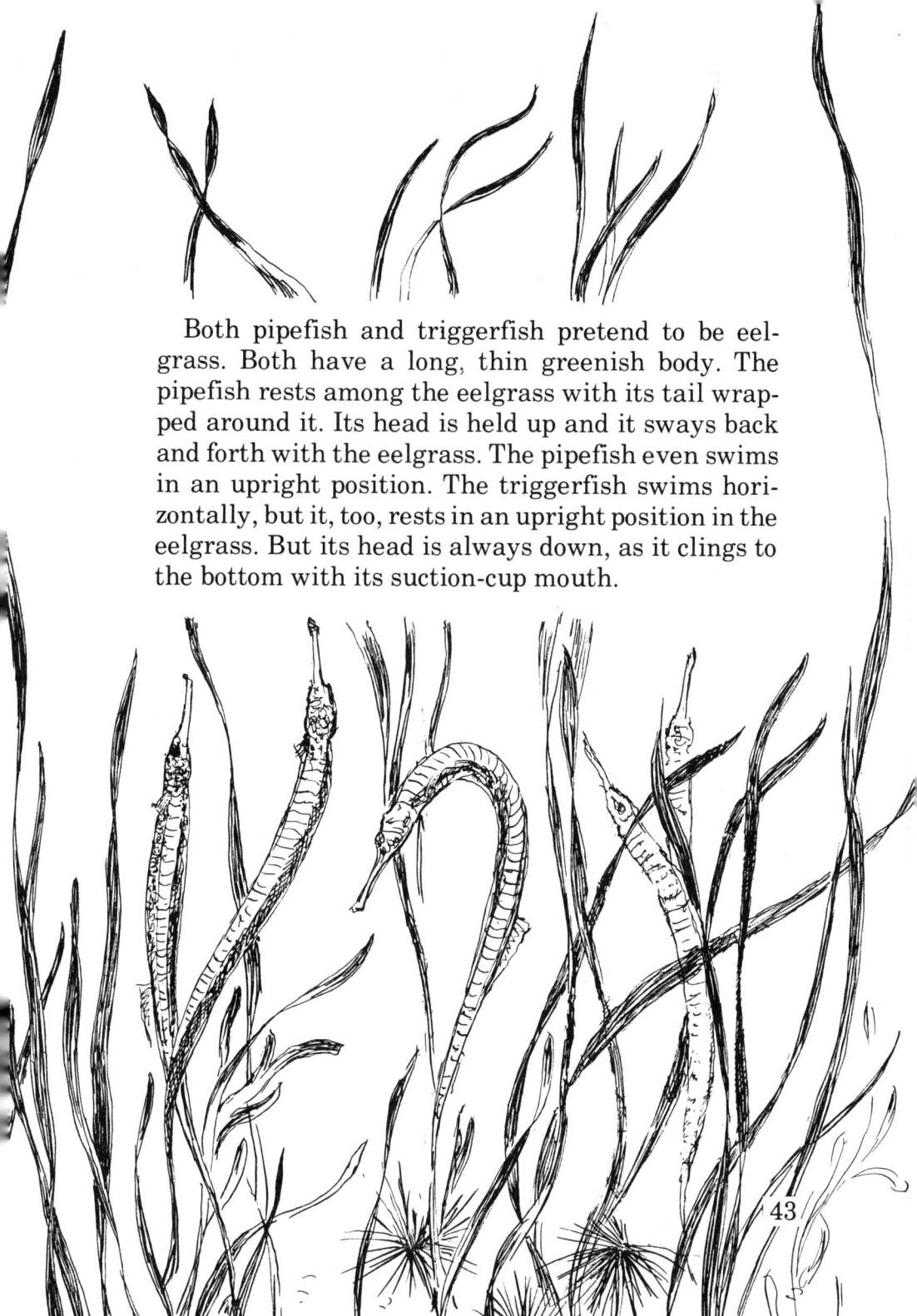

Both pipefish and triggerfish pretend to be eelgrass. Both have a long, thin greenish body. The pipefish rests among the eelgrass with its tail wrapped around it. Its head is held up and it sways back and forth with the eelgrass. The pipefish even swims in an upright position. The triggerfish swims horizontally, but it, too, rests in an upright position in the eelgrass. But its head is always down, as it clings to the bottom with its suction-cup mouth.

PRETEND FLOWERS

On Madagascar Island, near Africa there is a small pink sap-sucking insect, *Flatid,* (flat-tid) which passes for a flower petal when it alights on a plant. If frightened, it falls away and drifts down, looking like a falling flower petal.

Groups of *Flata,* a butterfly of East Africa, join together to resemble the flowers or buds of the foxglove plant. *Flata* are either red or green. When they alight on a plant, the green ones arrange themselves to look like unopened buds, and the red ones to look like red foxglove blossoms.

In Malaysia there is an amazing praying mantis, *Hymenopus coronatus,* (High-mee-nop-us koor-oh-nat-us) which looks like a pink orchid. It often hides among orchid blossoms, passing as one, while it waits to catch the insects which come looking for nectar. On a bare branch the mantis looks like a full-blown orchid and so escapes the lizards and birds who would like to eat it.

BORROWED DISGUISES

Some animals are not born with disguises. They find materials and make one.

Masking or decorator crabs are experts at disguising themselves with materials they find around them. Each species of decorator crabs disguises itself in a slightly different way. Some species cover only their heads, some their entire bodies, and some only the front legs. They use their large front claws to cut or tear out the right size pieces of seaweed and sponge. Then they chew on one end to make it sticky, and fasten the material to the hooklike bristles which cover them. Sometimes they add barnacles, sea anemones, coral, or moss animals—parasites which are already growing on the crabs' shells.

The young sponge crab cuts out a piece of sponge the size of its back. It holds this sponge over its back with a pair of hind legs used only for that purpose. If the sponge keeps on living and growing, the crab has a permanent disguise. If the sponge dies, the crab throws it away and cuts another.

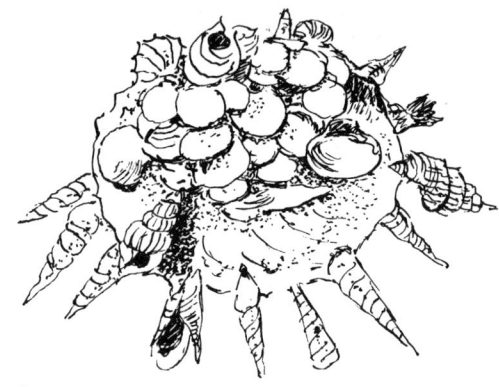

The *xenophora,* (zen-oh-for-ia) or carrier shell, cements dead seashells, pebbles, bits of coral and other debris to its shell. When small, the carrier shell uses small things for decoration. As the carrier shell gets larger, it uses larger pieces. But at all times it looks like a pile of rubbish which no animal would want to eat.

Insects also borrow disguises. The trash carrier feeds on aphids and other small insects. After it has sucked an insect dry, it fastens the remains onto the hooked bristles covering its body. It looks like a small moving trash pile as it crawls around covered with the remnants of past meals.

The caddis fly *larvae,* the early, wingless, form of the insect, live underwater in a case stuck together with glue from special glands. Each species builds a different type of case using different materials. They must make a new and larger case when they outgrow their old ones. Materials used can be twigs, leaves, bark, sand, tiny pebbles, and almost anything available in the pond or stream where they live.

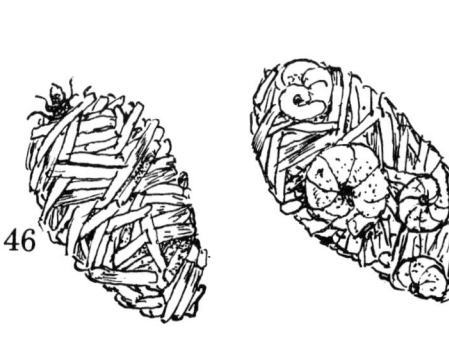

Chapter 6

Tricks That Fool

WRONG TARGET

Sometimes the body of an animal has markings or a growth which attract the eye of the predator who strikes at this part of the body. But these markings are on non-vital areas, not needed for survival. An animal can live without a tail, leg or piece of wing, but not a head.

Some butterflies and moths have *eyespots*, large fake eyes, on their wing tips. When a bird tries to catch an insect, it aims for what it thinks is the head. Butterfly collectors often capture peacock butterflies with V marks on the eyespots of their wings, the mark of a bird's beak.

One tropical butterfly of Central America has a fake head on its tail. Formed from parts of the wings, this false head is the shape and color of the butterfly's real head. There are spots for eyes, and even slender pieces of wing sticking out to look like antennae.

One little fish from New Guinea has large eyespots near its tail. It swims backward to keep up the disguise. If attacked, it confuses the predator by darting swiftly away in the opposite direction.

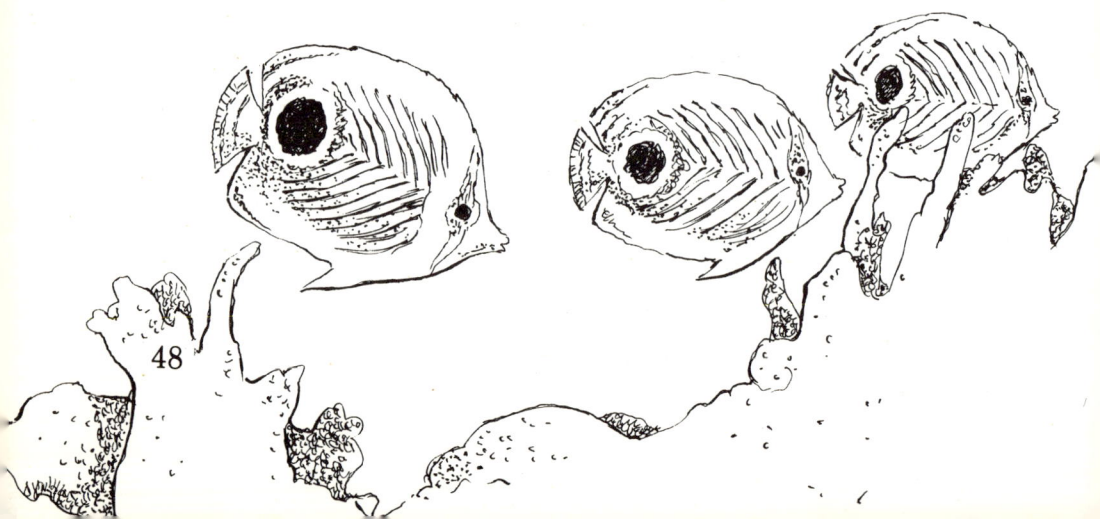

BREAK-AWAY TAILS

The tails of various lizards and salamanders break off easily but, in time, grow back. These lizards usually have brightly colored or marked tails to attract attention. The predator attacks, and the tail breaks, sometimes into several pieces. When this happens, the predator is confused, and the lizard has time to scurry away.

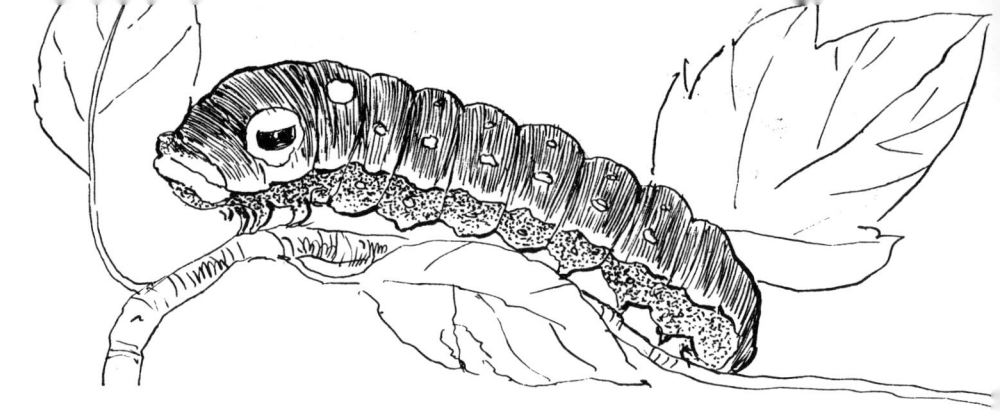

SCARING THE HUNTER

If an animal can make itself look dangerous, even though it isn't, it can sometimes frighten away a predator.

The larva of the spicebush swallowtail butterfly has a small head. The tail end, which is much larger, has a pair of black and yellow eyespots with nostril-like markings on it. The larva can puff up this end and these staring eyes usually scare any predator away.

The puss moth caterpillar usually looks quite harmless. But if frightened, it gets pretty scary. It raises both head and tail. One end has the markings of a horrible face on it, while the other has stingerlike horns. The sphinx moth caterpillar, although quite harmless, also scares away predators. When threatened, it rears back, raises its front legs, and exposes its eyespots there. The horn on its tail looks like a stinger.

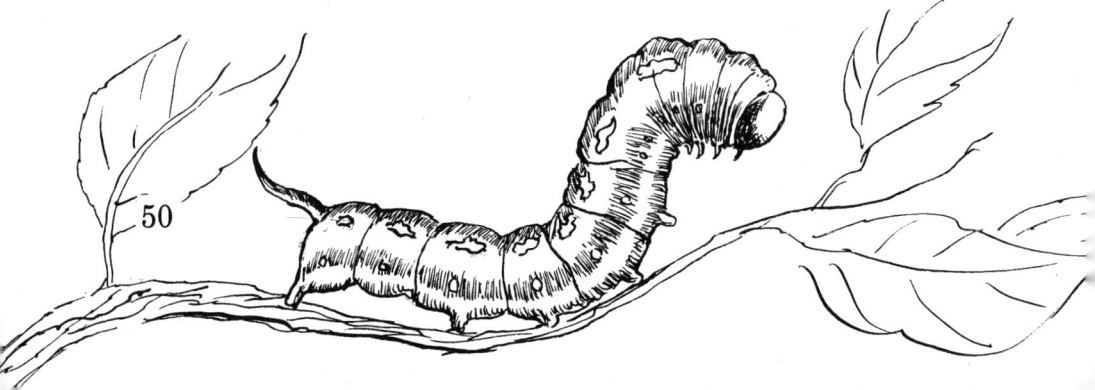

The harmless larvae of many insects use eyespots to try to look like snakes. Many of them sway from side to side with snakelike movements. One African caterpillar waves its front set of legs in imitation of the snake's flickering forked tongue.

The mottled gray and brown owl's-head butterfly has large eyespots on its wings. When it rests on a branch, its wings are spread out and the colors blend in with the tree. The eyespots look like an owl's eyes staring out from between the branches.

The Atlas moth of the East Indies has a remarkable resemblance to a cobra's hood when it is resting. The way the wings are held, their shape and markings combine to resemble the facelike markings found on the back of this deadly reptile's hood.

The *laternaria* fly of South America is commonly called the alligator bug. In front of its real face, it has a large, ugly snout with alligatorlike markings. There are false eyes, snarling lips and even long teeth. Although it is only seven inches long, its strange look scares other animals away. Perhaps this is because alligators are feared in the jungles where the alligator bug is found.

Some animals try to scare others by appearing to be larger than they are. Several species of toads, frogs, and chameleons gulp in air to puff up their bodies. The puffer fish inflates itself by gulping in water. If it has been caught and is out of the water, it gulps down air. Some species are also covered with spines which stick out when the puffer is inflated. No predator would try to eat this pin cushion.

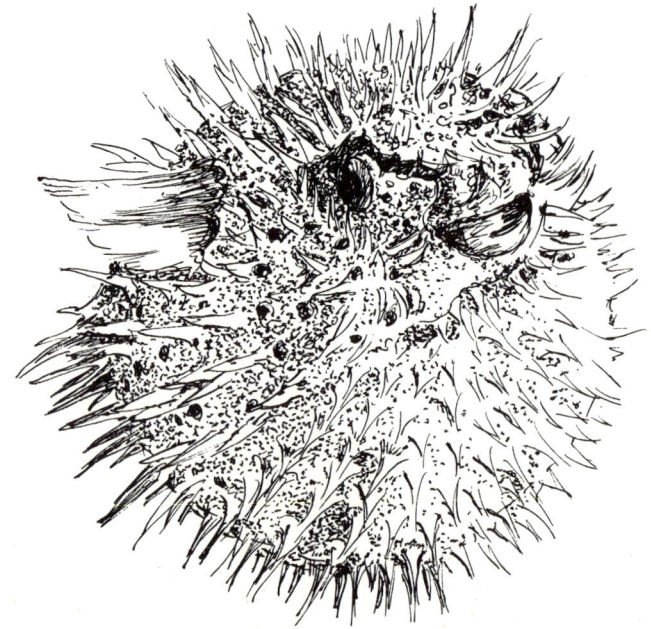

PLAYING DEAD

When all other methods of escape have failed, a few animals pretend to be dead. The opossum is the best known of these pretenders. We even use the term "playing possum" when someone is pretending to be asleep or unaware.

The opossum lies still and limp, its lips curled back in a grimace. Its heartbeat is slowed. It may really be in a state of shock caused by fright. Often the predator sniffs at it a few times and then goes away leaving the "dead animal" alone. After a few minutes, the opossum opens one eye. If the predator has gone, the opossum gets up and takes off.

The harmless hognose snake also puts on a good act when threatened. First, it tries to act like a dangerous animal. It flattens its head, hisses, and puffs itself up to look larger. If this fails, it shakes a few times and falls over, belly up, with its mouth open and plays dead. If turned over, it quickly flips over on its back again.

BROKEN WING ACT

Many birds which nest on the ground pretend to be hurt if a predator approaches too closely. This is done to protect the eggs or young. Usually one parent stays quietly on the nest. The other flutters away, crying pitifully. The predator thinks it will be easy to catch this bird with the broken wing and follows after. When the bird has led the animal away from the nest, it suddenly flies away, strong and whole.

Chapter 7

Animal Defenses And People

Humans have usually felt that large carnivores are harmful. They have tried to get rid of them. Wolves have all but been eliminated in the United States. The only large predators who still survive in the wild in the United States are coyotes and a few mountain lions.

The excuse for killing off predators is often that they kill other animals people feel are more desirable, or which they are raising for food. But these animals have evolved together over millions of years. Each has helped the other to survive.

The living things of the earth are bound together into a vast network. They are all dependent upon each other, and no animal can be considered alone. All animals are not completely protected by their defenses. Some are killed in spite of them. If a species were totally protected, other animals dependent upon it for food would die out. So, when a species improves its defenses, species that prey upon it improve their hunting skills.

Nor is it in the best interest of a species for all members to survive. Predators kill the animals whose defenses are weakest: the aged, the sick, and the weak. The food supply is then used by the stronger animals to make themselves even stronger. These strongest animals are also the ones left to reproduce, so the entire species becomes stronger.

What has happened in areas where all the predators of deer were killed off? Deer increased until there was not enough food for all. Often the hungry animals used up all their food supply so no more would grow. Instead of only the strongest surviving in good health, all survived in poor health. But many could not survive the winter months. They died off, and there were fewer deer than before.

In one area, hunters wanted all coyotes killed so there would be more quail. When all coyotes had been trapped, poisoned or shot, the quail also disappeared. Biologists found that the coyotes' prey had been the cotton rat. The rat's chief prey was the eggs and young of the quail. Without coyotes to control the cotton rats, more quail were killed than before.

It has taken millions of years for animals to evolve their systems of defense. But each has found a way for the species to survive. When humans remove one species, other species sometimes cannot survive. All animals have a place in nature—there are no good or bad ones. All fit together into a vast, balanced network. We do not have the right to decide which species should survive and which should be destroyed. Animal defenses work so all species will survive if we allow them to operate.

STUDYING ANIMAL DEFENSES

Do you want to learn more about the ways animals protect themselves? Watch them in your own backyard, city park or a vacant lot. The seashore teems with animal life. Especially watch the insects, as they have the greatest variety of defenses. Sit quietly, just watching, and maybe you will see a green stem move as a praying mantis grabs its prey. Perhaps you will see a "twig" come to life and crawl off.

Ask yourself questions about the animals you see. How do they live? What are the purposes of their different parts? Why is one animal one color and not another? Why do birds avoid certain insects? Take notes of the things you see. Maybe you will discover a new disguise, or facts about a disguise that were not known before.

Knowing about the way animals are protected and the adaptations which have helped them survive will help you understand some of the strange and wonderful things in nature. Most of our knowledge of animal behavior comes from people like you going out and studying animals where they live.

GLOSSARY

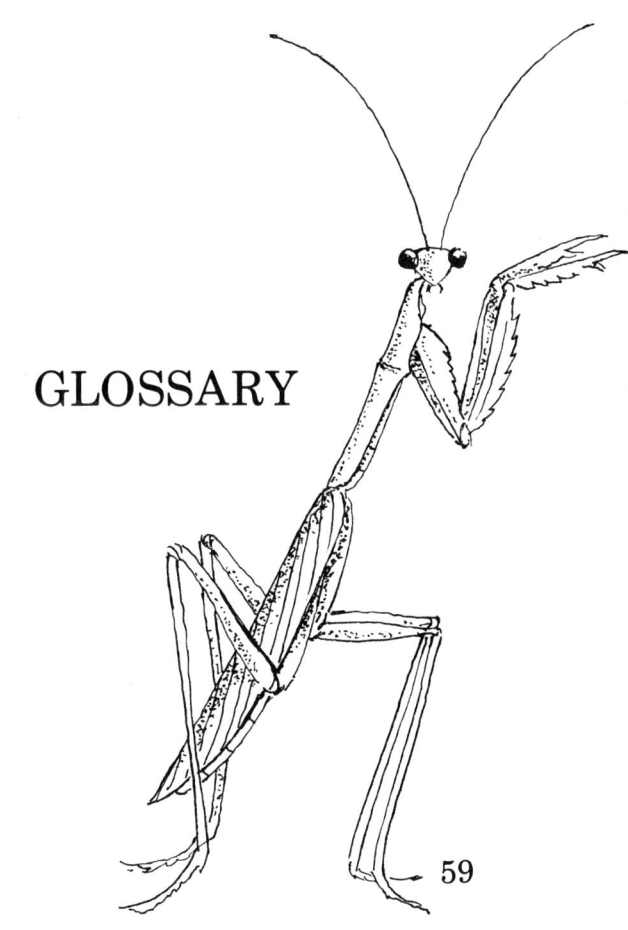

adapt   to change in a way that helps a living thing to survive

animal defenses   the ways animals protect themselves

camouflage   a disguise which makes something more difficult to see

canine teeth   the longer teeth on each side of the front ones in many carnivores

carnivores   meat-eating mammals

countershading   coloring, usually darker on the back and lighter on the sides and belly, which helps hide the rounded outline of an animal

defensive adaptation   the ways a species has changed to help it escape predators

disruptive coloring   markings which make it difficult to see the outlines of an animal

eyespots   markings which look like eyes

formic acid.   See venom

food chain   the links in the process by which animals eat plants or smaller animals and are in turn eaten by larger animals

fossil   hardened remains of a plant or animal from the distant past

habitat   the native living place of an animal or plant

inedible not suitable for food

larvae the immature stage of any animal without a backbone

mimicry the act of looking and acting like a different animal or object

"playing possum" pretending to be dead or asleep

polliwog or *tadpole* the young of a frog

predator an animal which preys on other animals

prey an animal hunted for food

primate an order of animals, including man, monkeys, and apes

protecting coloring colors or markings on an animal which help it hide

reverse countershading countershading when the belly is darker and the back lighter because the animal lives in an upside down position

species one kind of animal having common characteristics

tadpole. *See* Polliwog

tusks teeth developed to great lengths

venom a poisonous fluid produced by special glands in some animals and introduced into another animal's body

INDEX

alligators, 20, 26
alligator bugs, 52
antelopes, 11, 16
anole lizard, 23
ants, 13, 17, 33
aphids, 46
aquatic mammals, 12
armadillo, 20
armadillo lizard, 20
apes, 12
Atlas moth, 51

barnacles, 20
bats, 11
bears, 12
beavers, 12, 14
bees, 13, 17, 31, 33
beetles, 19, 33
birds, 12, 14, 23, 54
bobcat, 9
buffaloes, 16
butterfly, 35, 47-48

caddisfly larvae, 46
caribou, 16
carrier shell, 46
cat, 12
catfish, 20, 28
caterpillar, 28, 32, 40, 51
chameleon 29, 52
clams, 20
coral snake, 32
cotton rat, 57
coyotes, 55, 57
crab, 20
crab spider, 29
crocodiles, 20
cuttlefish, 18, 29

dead leaf butterfly, 36
decorator crabs, 45
deer, 11, 16, 22, 26, 56
di Cesnola, A. P., 24
electric catfish, 10
electric eel, 10
elephants, 15, 20

fish, 9, 12, 14, 26, 28, 29, 38, 48
Flata butterfly, 44
Flatid, 44
flies, 33
flounder, 29
flower flies, 33
food chain, 9
frog, 12, 22, 32, 52
frogfish, 41

gazelle, 11
geckos, 29
Gila monster, 32
giraffe, 25
goat, 12, 16
gold frog, 32
grasshopper, 19, 29, 33

harlequin cabbage beetle, 32
harp seal, 29
hermit crab, 20
heron, 9
hognose snake, 53
hornbill, 14
hornets, 13, 17
horse, 10, 16
hover fly, 33
hummingbird, 34
hummingbird moth, 34

62

inch worm, 40
insects, 11, 31, 58

jaguars, 25
jellyfish, 17

Kallima butterfly, 36
kangaroos, 11

laternaria fly, 52
leaf-cutting ants, 33
leaf fish, 38
leopard, 25
lizard, 20, 22, 26, 49
lobsters, 20

masking crabs, 45
monarch butterfly, 34
monkey, 12
moths, 33, 35, 47
mountain lion, 55
musk oxen, 13, 16

ocelot, 25
octopi, 18, 29
opossum, 53
ostrich, 11
oven bird, 14
owl's head butterfly, 51
Oxybelis snake, 40
oyster, 20

pangolin, 20
peacock butterfly, 47
peche de folha, 38
peppered moth, 30
pig, 15
pipefish, 43
polar bear, 12
pollywog, 9

porcupine, 20
praying mantis, 24, 39, 44, 58
primates, 12
puffer fish, 52
puss moth caterpillar, 50
puma, 26

quail, 57

rabbit, 11
reindeer, 16
reptiles, 12
rhinoceros, 20
roadrunner, 11

salamander, 17, 32, 49
sand grouse, 23
sargassum fish, 41
scorpion fish, 17
scorpions, 17
sea dragon, 42
sea horse, 42
sea robin, 20
sea snake, 17
serval, 25
sheep, 12
sloth, 22
snail, 20
snakes, 17, 22, 40
sphinx moth caterpillar, 50
spicebush swallowtail butterfly, 50
spider, 17, 28, 33
sponge crab, 45
squid, 18, 29
stingray, 17

tapir, Malayan, 26
tiger, 25

toad, 17, 32, 52
trash carrier, 46
trigger fish, 43
turtles, 20, 26

venom, 17
vervet monkey, 26
viceroy butterfly, 34

vine snakes, 40
walking-leaf beetle, 37
walking-stick insects, 39
walrus, 15
wasps, 31, 33
wolves, 55

zebra, 11, 25